ALPHAMETICS AS EXPRESSED IN RECREATIONAL MATHEMATICS MAGAZINE

Edited by

Charles Ashbacher

5530 Kacena Ave.

Marion, IA 52302

cashbacher@yahoo.com

Artwork

Caytie Ribble

Technical Advisor

Gisela Hausmann

ISBN-13 978-1508538134

ISBN-10 1508538131

Contents

Editor's Notes

I would like to thank Ray Madachy, the son of Joe Madachy, for giving permission for the use of this material from **Recreational Mathematics Magazine**. It is important that work like this be preserved and processed so that none of it is ever lost. Joe started **RMM** and was the editor and publisher for the fourteen issues it was published.

Caytie Ribble is the artist that created the original cartoon that appears on page 4. Joe Madachy included cartoons with a mathematical theme in **Recreational Mathematics Magazine** and he would have enjoyed this one as well.

Vulgar Fractions

An older term for a fraction that has integers in the numerator and denominator is "vulgar fraction."

Introduction

When Joseph (Joe) S. Madachy created **Recreational Mathematics Magazine (RMM)** in 1961, his goal was to give readers and subscribers the opportunity to solve problems that forced them to expend some mental energy. One of the staples of recreational mathematics from the beginning has been the alphametic, there has been a section devoted to alphametics from issue 1 of **RMM** published in 1961 to the last issue of **Journal of Recreational Mathematics Volume 38 Number 2** that was published in 2014.

The problem is a simple one, an arithmetic problem is altered by replacing all or nearly all of the digits with letters, it is convention to put all of the letters in uppercase. While most of the published problems are in base-10, other bases have been used. Of course the number of different possible letters that can be used is always identical to the size of the base as the correspondence between letters and digits is always kept 1-1.

As the reader will learn as they work through the pages of this book, alphametics were originally called cryptarithms. Like so many staples of puzzling, the famed English puzzle master H. E. Dudeney played a big role in introducing these problems to the public. He was the originator of the classic

```
SEND
MORE
――――
MONEY
```

addition alphametic.

The term "alphametic" was coined by J. A. H. Hunter to refer to cryptarithms where the letter replacements form words that make up an intelligible message. As you will see as you look through the pages of this book, Hunter was a very prolific creator of alphametic problems.

Solving the standard alphametic is something that all people from grade school on up can do. All that you need is the basic knowledge of arithmetic, a realization that a leading zero is not allowed and some logic and perseverance. That makes them ideal for brain exercises and when they are properly explained they can be used to teach the laws of arithmetic.

This book is a republication of the alphametics that were published in **Recreational Mathematics Magazine** during the fourteen issues that were published. In nearly all cases the problems are reproduced as they appeared in **RMM**, although the notes have been moved a bit to improve the format.

Brief historical notes were included in the alphametics sections of **RMM** and they have been included here, the only changes have been in their relative locations. From these notes you can learn some of the background and history of this valuable problem in recreational mathematics.

While this book is dedicated to the memory of Joe Madachy I would also like to commend Steven Kahan, the longtime editor of the alphametics column of **Journal of Recreational Mathematics**. His years of effort kept the column alive and with approximately ten new problems per issue four times a year for so many years, he has handled hundreds of problems with dedication and professionalism.

I encourage all readers to send their comments to me at cashbacher@yahoo.com.

SOLVING ADDITION ALPHAMETICS

Charles Ashbacher

Solving an addition alphametic is a process of logic that reduces the number of possibilities followed by trial and error, working through the remaining options. This section will be a brief explanation of how to solve them and it did not appear in **Recreational Mathematics Magazine**. The problem worked through will be the classic

```
 SEND
 MORE
_____

MONEY
```

While bases other than ten are occasionally used, if there is no base specified then the problem is to be solved in base-10. The first thing to note is that there are eight letters in the problem, so two of the digits will not be used.

Since leading digits generally cannot be zero, it is immediately known that S and M cannot be zero, Furthermore, D and E also cannot be zero, for if one was zero the other letter would be the sum of the ones position.

We then can quickly conclude that $M = 1$, for that is the largest possible carry out from the positional addition

```
carry in
   S
   M
 _____
```

This forces S to be either 8 or 9, with 9 the most likely, since

```
carry in
   S
   1
 _____
```

has to be at least 10. Since 1 has already been allocated, this forces O to be zero and allows us to conclude that $E + 1 = N$. Furthermore, we also know that

```
carry in
   N
   R
_____
```

is greater than 11 so that we have the carry in of one in the hundreds column and no zero or one for the column sum. Applying similar reasoning we can conclude that D + E has to be at least 12.

At this point we know enough so that we can try all the possible pairs of values for E and N,

E = 2, N = 3; E = 3, N = 4; E = 4, N = 5; E = 6, N = 7; E = 7, N = 8

and fairly quickly find the solution

```
   9567
   1085
_____
  10652
```

While this is a fairly simple addition alphametic, the basic principles of arithmetic and logic along with some perseverance will allow you to solve all basic alphametics. The growing power and ubiquity of computers has led to a dramatic increase in the size of alphametics that have been created. The alphametics column of volume 36, No. 1 of **Journal of Recreational Mathematics** that was published in 2007 contained a problem that has 24 summands, putting it beyond the reach of the solver restricted to pencil and paper.

THE ALPHAMETICS THAT APPEARED IN RECREATIONAL MATHEMATICS MAGAZINE

Issue Number 1, February, 1961

Alphametics is the name applied to mathematical puzzles in which most or all of the digits are replaced by letters. The name was invented by J. A. H. Hunter and is used by him, and others, to refer specifically to those puzzles in which the combinations of letters used make sense.

Mr. Hunter has devised a couple of alphametics especially for the first issue of **Recreational Mathematics Magazine**.

This is a rather easy alphametic. Each letter stands for a particular, but different figure. What do you make of our FUN?

```
    RMM
    OUR
    RMM
    OUR
    RMM
   ─────
    FUN
```

This is a really difficult alphametic. Each letter stands for a particular, but different, figure and the little crosses indicate figures about which we know nothing. What do you think of the FIRST RMM?

```
THE ) FIRST ( RMM
      xxxx
     ─────
      xxxx
      TOUR
     ─────
       xxxx
       xxxx
      ─────
        xM
```

Gerald Mosler of Long Island sends in the following alphametics for your further enjoyment – or frustration. The same rules apply as outlined by Mr. Hunter above.

Here's a breakfast order which, when added up as shown, may not be to your liking.

```
    HAM
   EGGS
   ────
   MASH
```

When these two get together there's certain to be much ado about something!

```
   DOG
   CAT
   ───
   ADO
```

Now let's see if you can add animals together as shown here. Darwinists won't go along with this, we're sure.

```
   BEAVER
    TIGER
   ──────
   RABBIT
```

Alphametics

Issue number 2, April, 1961

In May 1931 the use of the word *Cryptarithmie* (Cryptarithmetic) appeared in the magazine **Sphinx**. A *cryptarithm* is a mathematics problem involving addition, subtraction, multiplication, or division in which the digits have been replaced by letters of the alphabet or some other symbols. The term *alphametics*, originated by J. A. H. Hunter, is used to refer specifically to those *cryptarithms* in which the combination of letters make sense.

Here we offer some alphametics for your fun.

Young and old, they all enjoy these! Each letter here stands for a different figure, and little crosses indicate figures about which we're told nothing. What do you make of the FOOLS? (J. A. H. Hunter)

```
ALL ) FOOLS ( DAY
        Gxx
       ____
       xAxx
       xxMx
       ____
        xxEx
        xxxS
        ____
```

In keeping with the reputation of the magazine, as implied by the wording, this is quite a tricky alphametic. The same rules apply as explained above. What do you make of the GAMES? (J. A. H. Hunter)

```
        RMM
        SLY
       ____
        xxxx
       xxRx
       xxM
       ____
       GAMES
```

A little game of poker was played the other night and the originator of the multiplication alphametic below tells us that the card filling the winning hand is represented by the letter "A". (Mrs. Rae C Nelson)

```
        (A)(SPADE) = FLUSH
```

Alphametics

Issue number 3, June, 1961

This certainly won't be easy! But it can be done. (J. A. H. Hunter)

```
       MOON
        MEN
        CAN
       ____
       REACH
```

A vexing problem, we're sure! (G. Mosler)

```
(V)(VEXATION) = EEEEEEEE
```

As Shakespeare very nearly wrote:

```
ALLS
WELL
THAT      But, we ask, what's SWELL if you have no one? (D. Murdoch)
ENDS
─────
SWELL
```

Et voila une aphametique! Every French student will be able to agree that the addition of one, one, two and twelve amounts to sixteen – but what different figure does each letter represent if SEIZE is itself very properly divisible by 16? (D. Murdoch)

```
   UN
   UN
 DEUX
DOUZE
─────
SEIZE
```

Alphametics

Issue 4, August, 1961

Alphametics have been around for quite awhile and the editor has had the two given here sent to him several times. The origin of each is unknown – but many **RMM** readers may not have seen them, or may have forgotten the answers if they have seen them.

```
SEND                EVE
MORE                ───  =  .TALKTALKTALK . . .
─────               DID
MONEY
```

Now the English may not be perfect, but this little alphametic has a unique answer. (Alan Gold)

```
  CATS
   EAT
 ─────
 xxxxx
 xxxxx
xxxxx
 ─────
xMOUSES
```

Alphametics do not necessarily have to be based on our usual decimal number system. So what if the value of FARMER expressed in the lowest-base system applicable here? (Margaret M. Rohe)

```
WHEAT
FIELD
———————
FARMER
```

Dan and Edna may have had a good time at Aden, but what kind of a time will you have trying to solve this alphametic? (A. G. Bradbury)

```
 DAN
 AND
EDNA
  AT
————
ADEN
```

Here's an alphametic in a rather different form. It's a bit difficult so be wary (J. A. H. Hunter)

$$\sqrt{CAREER} \quad = NOW$$

Alphametics

Issue 5, October, 1961

It is the intent to publish alphametics with unique answers. However, sharp **RMM** readers sometimes find two or more answers to these puzzles – and we like this!

This may be a very appropriately-worded little puzzle! (J. A. H. Hunter)

```
THESE
TEASE
TRIED
——————
READER
```

(Afterthought by the editor: A year's subscription, or extension, to **RMM** for anyone who finds a solution to this Alphametic if the word TIRED is substituted for TRIED.)

This alphametic is for the BIRDS! But try to solve it anyway. (Alan Gold)

```
    SEE
    THE
    ─────
   Bxxx
    Ixx
  xxRx
  ────────
  xxxDSx
```

When Edward, the big game hunter, followed the trail of the man-eating tiger a shade too far into the jungle, the local reporter filed the following cable to the news agencies:

```
ED : IN : : DEN : END
```

(It will be noted that the punctuation is somewhat unconventional – though mathematical.) What information can be DENIED? (Derrick Murdoch)

 BBG for Canadians needs no explanation – but for U. S. readers of **RMM** it means Board of Broadcast Governors and this alphametic is appropriately worded. (Alan Gold)

```
    OH
    NO
   ─────
   NEW
   BBG
   ─────
   BLOW
```

Alphametics

Issue 6, December 1961

   The interest of **RMM** readers in this department prompts the editor to present a bit of history of the type of puzzles found here.

 The very first "hidden digit" problem will probably never be known, but it is easy enough to go back to 1906 when W. E. H. Berwick presented a division problem of 79 digits with only seven sevens given as clues. (**School World Vol. VIII** pp. 280, 320). Prof. Schuh of Delft in 1921, managed to devise a long division problem in which *none* of the digits were given directly as clues. These "hidden digit" problems did not utilize letters or symbols to match particular digits – blank spaces were left to be filled in. **Strand Magazine**, in 1921, published a "hidden digit" problem using letters of the alphabet and, later, some problems with chessmen covering the digits appeared. The word *cryptarithmie* (cryptarithmetic), attributed to Minos, was used in the May 1931 issue of **Sphinx**.

So far, any letters or symbols could be used in place of the digits. However, letter substitutions which formed words and even phrases proved more popular. A truly ideal cryptarithm is one in which all the digits from 0 through 9 are substituted by letters which form a sensible phrase and in which only *one* solution is possible.

J. A. H. Hunter, who devised the term alphametic to apply to a cryptarithm in which the letter substitutions make sense, first used the term *alphametic* in the **Toronto Globe & Mail** of October 27, 1955. The term is credited to a typographical error made by one of Mr. Hunter's correspondents at that time.

Here it is, the first cryptarithm published under the name *alphametic*. We leave it to be solved by **RMM** readers.

```
BE ) ABLE ( SIR
       MR
      ─────
      RRL
      RLM
      ─────
       BE
       BE
```

Each letter in this little message stands for a different figure, and we trust you will indeed make MERRY. (J. A. H. Hunter)

```
 HERES
 MERRY
  XMAS
 ──────
READER
```

By now we have heard Bong Crosby sing *White Christmas* too many times – but this is a time for joy, so let's SING SONGS. (A. G. Bradbury)

```
   DO
 SING
 BING
 SING
 ─────
SONGS
```

A true addition certainly! And a merry TWELVE days of Christmas to all **RMM** readers who solve this little gem of an aphametic. (Herman Ebbers)

```
    TWO
  THREE
  SEVEN
 _____

  TWELVE
```

Alphametics

Issue 7 February, 1962

A little note to be added to the brief historical sketch in the December, 1961 **RMM**: the **Strand Magazine** for July 1924 published the following alphametic:

```
(TWO)(TWO)  =  THREE.
```

J. A. H. Hunter, Maxey Brooks and the editor (and, undoubtedly, many other **RMM** readers) are still trying to locate the earliest published alphametic (a cryptarithm with meaningful words). Also, when did the classic SEND + MORE = MONEY first appear? Any help by **RMM** readers would be appreciated by all.

This is the first issue of the year, so let's TRY AN EASY ONE. (A. G. Bradbury; North Bay, Ontario)

```
AN ) EASY ( ONE
     xT
    _____

    xxx
    xxR
    _____

      xx
      xY

      _____
```

"Is that new?" Tony asked, as he strolled through the zoo. And the keeper replied: "It's not new, it's a gnu." (J. A. H. Hunter; Toronto, Ontario)

```
BIG ) NUGS (OR
       BIG
      ─────
      GNUS
      GNUS
      ─────
```

After two such easy alphametics, see how well you do with the following two.

This is a wild one all right! How about taming the TIGERS? (Alan Gold; Downsview, Ontario)

```
    SEE
    THE
   ─────
   xxxx
  xxSx
 xxxH
  ─────
 TIGERS
```

Mr. Bradbury gives us another poser. Just don't get seasick on this one! (A. G. Bradbury; N. Bay, Ontario)

```
YOHO ) HEAVE ( HO
       xxxx
       ─────
        xxxx
        xxxx
        ─────
          UP
```

Alphametics

Issue 8, April, 1962

With Col. Glenn's successful orbital flight we feel this little alphametic is quite appropriate! (A. G. Bradbury; N. Bay, Ontario)

```
    TRACK
    SPACE
   ──────
   ROCKET
```

This division alphametic should keep all **RMM** puzzle-solvers testing out quite a few numbers. (A. G. Bradbury; N. Bay, Ontario)

```
TWO ) TWELVE ( SIX
        xxxx
        ————
        TExx
        xSTx
        ————
         OUx
         xTx
         ———
          xx
```

There's obviously no mistake in this little addition. So, one way or another, you should have no difficulty in evaluating that EIGHT. (Harvey Hahn; Valparaiso, Indiana)

```
          ONE
          TWO
         FIVE
        —————
        EIGHT
```

Even a biologist won't find this easy! What is this BEETLE? (Anneliese Zimmerman; Montreal, Quebec)

```
          FEED
           THE
        ——————
        xxxxx
        TESTx
        xxxx
        ——————
        BEETLE
```

There are very few completely hidden cryptarithms and we offer this one even though it is not an alphametic. (Nigel Mason; Liverpool, England)
```
```

```
xx) xxxx (xx.xxx
    xx
    ___
    xxx
     xx
     ___
      xx
      xx
      ___
       xxx
       xxx
       ___
        xx
        xx
        ___
```

Alphametics

Issue 9, June, 1962

 Research into the origin of the classic SEND MORE MONEY message has revealed that it was not an appeal to his dad by an impoverished college boy. It must have been written by some girl, for the father sent along all his spare cash with the reply that has now come to light. What is the largest amount of CASH he can have sent? (W. A. Robb; Ottawa, Ontario)

```
ALAS
LASS
  NO
MORE
____
CASH
```

This little alphametic is surely seasonable. What is your idea of WARMER? (J. A. H. Hunter; Toronto, Ontario)

```
  AIR
  AIR
 ____
  XXX
 XXXX
XXXX
_____
WARMER
```

The summer months mean swimming and other outdoor activities. Only those in top form will be able to dive into this ACT. (A. G. Bradbury; North Bay, Ontario)

```
HIGH ) DIVING ( ACT
         xAxS
        _____
         xxxx
         Txxx
        _____
         xxHxx
         xxxIx
        _____
          xxS
```

"Spring has sprung, and the birds are on the wing" as the saying goes. But some of the BIRDS are in FIRS. (A. G. Bradbury; North Bay, Ontario)

```
      FIRS
      FOR
      THE
     _____
     BIRDS
```

Alphametics

Issue 10, August, 1962

When this issue was delayed due to the editor moving from Idaho to Ohio, the following alphametic was put on the cover.

```
   PLEASE
   PARDON
  _____
   DELAYS
```

H. E. Dudeney, one of the greatest puzzilists of all time, seems to have been the originator of the classic alphametic SEND MORE MONEY. J. R. Whalley of Sussex, England sends in the information that in **Strand Magazine** for July, 1924, Dudeney, under "Perplexities", composed four alphametics. They include the classic SEND MORE MONEY and
(TWO) (TWO) = THREE previously quoted in the February, 1962 issue of **RMM** on page 13.

Here are Dudeney's alphametics which he entitled "Verbal Arithmetic":

Addition is an imposition

```
    SEND
    MORE
   ──────
   MONEY
```

Subtraction is as bad

```
   EIGHT
    FIVE
   ──────
   FOUR
```

Multiplication is vexation

```
    TWO
    TWO
  ──────
 THREE
```

Division drives me mad

```
TWO ) SEVEN ( TWO
       BOB
      ──────
       JOE
       OVV
      ──────
        VESN
        VESN
       ──────
```

The answers to Dudeney's alphametics will be given in the December, 1962 issue of **RMM**. We cannot guarantee that all the above have unique solutions.

Now, the only alphametic published in **RMM** whose authorship remains unknown is:

```
   EVE
   ─────  = .TALKTALKTALK . . .
   DID
```

To complete this issue's page of alphametics, we present two regular addition problems.

The sign seems to be set high at the bus stop. What do you make of that STOP? (A. G. Bradbury; North Bay, Ontario)

```
   POST
   TOPS
   ──────
   STOP
```

Teamwork is the modern battle cry. You may require help in solving this timely alphametic or it has TWIN solutions. (J. A. H. Hunter; Toronto, Ontario and Joseph S. Madachy; Kent, Ohio)

```
  NOW
  TWO
   IN
 TWIN
 ─────
ORBIT
```

Alphametics

Issue 11, October, 1962

This department appears to be the most popular feature in **RMM**. The Puzzles and Problems are indeed very popular too. However, we believe alphametics have greater appeal since no more than knowledge of the basic principles of arithmetic is required to solve the problems – plus the ability to apply these principles logically.

Perhaps other readers can supply their own reasons why such an essentially simple type of puzzle can appeal to – and stump – solvers from grade school level through post graduate college students.

We all know what this means. (H. S. Tribe; Sutton, England)

```
 NOEL
 NOEL
 ────
BELLS
```

Simple Simon's pseudo-summing certainly seems silly. Still, certain substitution solves Simple Simon's summing. (Robert Ruderman; New Hyde Park, New York)

```
 THREE
 SEVEN
  NINE
 ─────
TWELVE
```

Square dancing, as we all know, soon has us going in circles. This alphametic may keep you going in circles if you don't solve the hidden problem first. (A. G. Bradbury; North Bay, Ontario)

```
SQUARE
 DANCE
 ─────
DANCER
```

A colorful alphametic most assuredly! One that may turn some of you purple trying to solve it and make you green with envy seeing others solve it with ease. (A. G. Bradbury; North Bay, Ontario)

```
 GREEN
  GREY
  FAWN
 _____
YELLOW
```

North is north and south is south, so there can be no argument about the relative positions here. (R. Robinson Rowel Sacramento, California)

```
CANADA
UNITED
_____
STATES
```

Here's wishing all our readers a very happy New Year. (J. A. H. Hunter)

```
HAPPY
HAPPY
HAPPY
  DAYS
_____
AHEAD
```

Alphametics

Issue 12, December, 1962

Successful meetings are anticipated since we will make the CLUB as large as possible. (Donval R. Simpson, College, Alaska)

```
 MATH
 CLUB
_____
MEETS
```

Readers of **RMM** are likely to be MAD readers and ALFRED E. NEUMAN should be familiar to many. This mad little alphametic is in scale-9 notation. (C. R. J. Singleton; Sheffield, England)

$$ALFRED \div E = NEUMAN$$

There can be no argument about the total here. So what do you make of THREE? (Richard L. Breisch; Royersford, Pennsylvania)

```
    ZERO
     ONE
     TWO
  _____

   THREE
```

One of the reasons for the late issues these last few months. (Paul M. Nemecek; Riverside, Illinois)

```
      RMM
      HAS
   _____

     XXXX
     XXXX
      RMM
   _____

    MOVED
```

Although not quite an alphametic – unless it be taken as a result of alcoholic ebullience mixed with a touch of the sun – this cryptarithmic monster certainly has its points. For this interesting puzzle we are indebted to Steven R. Conrad of Clayton, Missouri.

```
        SUN
       LOSE
      UNTIE
     BOTTLE
    ELISION
   NINETEEN
  NONENTITY
  EBULLIENT
  INSOLUBLE
  _____

  NEBULOSITY
```

Alphametics

Issue 13, February, 1963

With this issue, **RMM** has published a total of 66 different alphametics and cryptarithms. The score by country: Canada 39 ½; USA - 15½; England - 9; Denmark – 1; Unknown – 1. In the

June, 1962 issue of **RMM** we tried to prod US puzzlists to catch up to Canada. The score, then, was 75% from Canada and only 16% from the US. Now, it's 60% and 24% respectively. Can we hope for more representation from other countries?

Our ANTS, of course, are obviously very odd. (Jonathan Khuner; Berkeley, California)

```
  ANTS
  CANT
  ————
  SCAN
```

In this unusual alphametic each separate column, including its digit in the final total, adds up to a sum that is divisible by 8. For example, the sum of E, A, A, and T is so divisible. (George Propper; Bronx, New York)

```
  OBEY
   SAM
   SAM
  ————
  VOTE
```

Here, we have a detailed calculation of the cube of a 4-digit number, in two separate operations. In the first operation we derive the square of the number; in the second we multiply the 4-digit number by its square. The little x's indicate the positions of the digits. What is the original 4-digit number? For this original and most intriguing example of the rare "no digits" form of puzzle, we are indebted to Willy Enggren of Copenhagen, Denmark.

```
     XXXX                    XXXX
     XXXX                  XXXXXXX
    ———————                ————————
    XXXXX                    XXXX
     XXXX                    XXXX
    XXXX                     XXXX
   XXXX                      XXXX
   ———————                  XXXXX
   XXXXXXX                  XXXXX
                          ——————————
                          XXXXXXXXXX
```

There can be no doubts as to what we have here. So what is this PRIME? (A. G. Bradbury; North Bay, Ontario)

```
     THIS
     SURE
       IS
     _____
    PRIME
```

The wise ones "give up" in time, but you don't have to here! (A. G. Bradbury; North Bay, Ontario)

```
    FLAT
      AS
    _____
    xxxx
    xOLD
    _____
   ACTOR
```

Alphametics

Issue 14, Jan. – Feb., 1964

In this alphametic someone has told Bob to go, but he says no. (Alan Gold; Downsview, Ontario)

```
OH ) BLOW ( GO
       BOB
      _____
       xxx
       xxx
      _____
        NO
```

On TV, maybe? But WHERE? (Harry L. Nelson; Livermore, California)

```
            CAR
        _____
  54 )  WHERE
        xxx
        _____
          xxx
          xRU
        _____
            xxx
            xxx
          _____
            ---
```

You don't have to be a lawyer to get what CAUSA means in this famous phrase:

```
CAUSA = SINE + QUA + NON
```

(Jonathan Khuner; Berkeley, California)

There are the experts, of course, but there is really nothing at all difficult about our GAMES.
(Jonathan Khuner; Berkeley, California

```
        BASE
        BALL
        _____
        GAMES
```

This is obviously taken right out of context, but it's quite easy to see what EVEN means. (Fr. Victor Feser; Richardton, North Dakota)

```
        NO
        IF
        _____
        xON
        xIF
        _____
        EVEN
```

Here's a true addition, where even TWO and TWENTY are even! (C. R. Dickinson; Camas, Washington)

```
   TWO
 SEVEN
ELEVEN
_____

TWENTY
```

This should be no challenge to the experts. (A. G. Bradbury; North Bay, Ontario)

```
EAGER
  SEA
EAGLE
RIDES
_____

GALES
```

We certainly see more in this issue. (J. A. H. Hunter; Toronto, Ontario)

```
SEE
RMM
SEE
____

MORE
```

SOLUTIONS TO ALPHAMETICS

Issue Number 1, February, 1961.

```
    RMM              155
    OUR              241
    RMM              155
    OUR              241
    RMM              155
    _____          _____
    FUN              947
```

```
THE ) FIRST ( RMM        178 ) 53241 ( 299
      xxx                       356
     _____                    _____
      xxxx                      1764
      TOUR                      1602
     _____                    _____
       xxxx                      1621
       xxxx                      1602
      _____                    _____
         xM                        19
```

```
    HAM          932           763
    EGGS        1447    or    2884
    ____        ____          ____
    MASH        2379          3647
```

```
    DOG    123      125      134
    CAT    689      687      579
    ___    ___      ___      ___
    ADO    812      812      713
```

```
    BEAVER    251453
     TIGER     60753
    _____    _____
    RABBIT    312206
```

Issue number 2, April, 1961

```
ALL ) FOOLS ( DAY          388 ) 91180 ( 235
      Gxx                        776
      ─────                      ───
      xAxx                       1358
      xxMx                       1164
      ─────                      ────
       xxEx                       1940
       xxxS                       1940
       ────                       ────
```

```
       RMM                   633
       SLY                   127
      ─────                  ───
       xxxx                  4431
       xxRx                  1226
       xxM                    633
      ─────                  ───
       GAMES                 80391
```

```
(A)(SPADE) = FLUSH      (5)(13582) = 67910
```

Since the statement given says that the winning hand (a FLUSH in this case) was filled by a card represented by A, then the total of 6 letters in A FLUSH must represent only five cards and, therefore, one of the letters in FLUSH must be a one (A cannot be 1). The only other solution fulfilling this condition is (4)(17453) = 69812, but this would mean a Queen by the notation 12 which is seldom done.

Issue number 3, June, 1961

```
       MOON        9552
       MEN          902
       CAN          382
      ──────      ─────
       REACH       10836
```

```
(V)(VEXATION) = EEEEEEEE     (9)(98765432) = 888888888
```

```
 ALLS     9332            9332
 WELL     8433            8433
 THAT     6596    or      6096
 ENDS     4072            4572
         _____          _____

SWELL    28433           28433


  UN         35             34
  UN         35             34
DEUX       8230    or     8632
DOUZE     84372          87316
         _____          _____

SEIZE    92672           96016
```

Issue Number 4, August, 1961

```
SEND        9567
MORE        1085
          _____

MONEY      10652
```

$$\frac{EVE}{DID} = .TALKTALKTALK \ . \ . \ .$$

$$\frac{212}{606} = .34983498 \ . \ . \ .$$

$$\frac{242}{303} = .79867986 \ . \ . \ .$$

```
       CATS         3462              3470
        EAT          546               947
                 _____          _____

      XXXXX        20772             24290
      XXXXX        13848             13880
      XXXXX        17310             31230
                 _____          _____

    xMOUSES      1890252           3286090

      WHEAT        95307             97305
      FIELD        1χ328            1832χ
                 _____          _____

     FARMER       104634            104634
```

Where χ stands for 10 in the base-11 system. H – I and T - D are interchangeable.

```
      DAN          542
      AND          425
     EDNA         3524
       AT           41
                 _____

     ADEN         4532
```

$$\sqrt{CAREER} = NOW \qquad \sqrt{561001} = 749$$

$$\sqrt{531441} = 729$$

Issue Number 5, October, 1961

```
     THESE
     TEASE          53646 + 56046 + 51269 = 160961
     TRIED
                 _____

     READER
```

```
      SEE          255              344
      THE          835              524
              _____         _____
     Bxxx         1275             1265
      Ixx          765              688      (A base-11 solution)
     xxRx         2040             15χ9
              _____         _____
   xxxDSx       212925           167935
```

```
ED:IN  ::  DEN:END        87:93  ::  783:837
```

```
       OH             58
       NO             25
            _____       _____
       NEW            290
       BBG            116
            _____       _____
      BLOW           1450
```

The TIRED variation on alphametic No. 1 was shown by many readers to be impossible in base-10 unless R = 0. Generally, alphametics are constructed so that the initial letter of a word is not equal to zero, unless some condition is specified. Since the Editor did not specifically eliminate this possibility the free 1-year subscription (or extension) is given to **RMM** readers who showed that no solution exists in base-10 unless R = 0. The first solutions in base-11 and base-12 are the ones the Editor was after.

```
THESE  +  TEASE  +  TIRED  =  READER

29747  +  27847  +  21076  =  078670  Base-10

18454  +  14954  +  17043  =  049340   Base-10

89353  +  83153  +  80237  =  231732   Base-11

90656  +  96356  +  97261  =  263162   Base-11

64757  +  67257  +  6317Ɛ  =  172Ɛ71   Base-12

72χƐχ  +  7χ4Ɛχ  +731χ5  =  1χ45χ1     Base-12
```

As a joke, the Editor challenged Mr. Carmony (who found all the solutions listed on the previous page) to find a solution in Base-119344307. He did! All 59,672,141 of them! He, too gets a lifetime subscription. (The Editor has stopped joking!)

Mrs. Rae Nelson of Livermore, California has submitted what amounts to an infinite number of solutions to the TIRED alphametic which earns her a lifetime subscription to **RMM** (since the offer was 1 year for each solution, Mrs. Nelson is obviously, getting gypped!)

The solution in any Base-L:

		T	H	1	L/2	1
THESE		T	H	1	L/2	1
TEASE		T	1	(L − 4)	L/2	1
TIRED		T	I	2	1	0
READER	2	1	(L − 4)	0	1	2

where L = 6k + 4, (k = 2, 3, 4, 5, . . .), H + I = L − 6 and T = (2L + 1) / 3. (If k = 0 then L = 10 and it has been shown that a solutionin Base-10 is possible only if R = 0, which is not allowed.) H and I are limited to values not assigned to the other letters.

For example, a solution in Base-28 yields T = 19, H + I = 22 and the following set-up (Base-10 notation is used to represent single digits in Base-28.)

```
   19  17   1  14   1
   19   1  24  14   1
   19   5   2   1   0
   _____
2   1  24   0   1   2
```

By using the indicated procedure, readers can quickly form solutions in Base-16, 22, 34, 40. etc., etc. Considering the above, the Editor, of course, closes the "contest" though he would be very happy to see any other unique developments.

Issue Number 6, December, 1961

BE   ABLE   SIR = 27 9207 341

HERES + MERRY + XMAS = READER

64143 + 74115 + 9783 = 148041

χ212Ɛ + 42113 + 549Ɛ = 129721 (Base-12)

52123 + 82116 + 4893 = 129021 (Base-11)

(χ = 10, Ɛ = 11 in the base-12 system.)

DO + SING + BING + SING = SONGS

83 + 1726 + 9726 +1726 = 13261

```
20 + 139(30) + (87)39(30) + 139(30) = 109(30)1 (Base-89)
```

And a multitude of solutions in other bases by Harry L. Nelson and J. A. H. Hunter.

```
TWO + THREE + SEVEN = TWELVE

106 + 19722 + 82524 = 102352

108 + 1Ɛ977 + χ7374 = 107237 (Base-12)
```

(χ = 10, $\mathcal{E}$ = 11 in the base-12 system.)

Issue Number 7, February, 1962

```
TRY AN EASY ONE = 640 18 5130 285

BIG NUGS OR BIG GNUS = 423 8037 19 423 3807

SEE THE SHE TIGERS = 933 263 963 245379

YOHO HEAVE HO UP = 1232 39489 32 65
```

Issue Number 8, April, 1962

```
TRACK SPACE ROCKET = 21469 85463 106932

TWELVE TWO SIX = 360410 367 982

ONE TWO FIVE EIGHT =621 847 9071 10538

FEED THE TEST BEETLE = 3446 274 2412 933204

xxxx ÷ xx = xx.xxx   1062 ÷ 16 = 66.375
```

Issue Number 9, June, 1962

```
ALAS LASS NO MORE CASH = 5157 1577 38 2804 9576
```

(CASH was to have been as large as possible.)

```
AIR AIR WARMER = 491 491 241081
```

(Many **RMM** readers turned in 496 496 246016 as an answer to this alphametic. We refer them to the actual set-up shown in the June, 1962 issue which will point out their error.)

```
HIGH DIVING ACT = 2102 917150 436
```

```
FIRS FOR THE BIRDS = 9063 976 584 10623
```

Issue Number 10, August, 1962

```
SEND MORE MONEY = 9567 1085 10652
```

```
EIGHT FIVE FOUR = 12780 6321 6549
```

```
TWO TWO THREE = 138 138 19044
```

```
TWO SEVEN TWO BOB JOE = 237 56169 237 474 876
```

```
EVE DID .TALKTALKTALK - 212 606 .349834983498 . . .
```

```
= 242 303 .798679867986 . . .
```

```
POST TOPS STOP = 3285 5238 8523
```

```
NOW TWO IN TWIN ORBIT = 312 921 73 9273 10579
```

= 214 941 52 9452 10659

Cover alphametic:

```
PLEASE PARDON DELAYS = 451681 463927 915608
```

Issue Number 11, October, 1962

```
NOEL BELLS = 9387 18774
```

```
THREE SEVEN NINE TWELVE = 13244 84546 6864 104754
```

It was pointed out by Harry L. Nelson of Livermore, California that $3 + 7 + 9 = 12$ in base-17 also – and he submitted all 340 distinct solutions in base – 17, e. g.:

```
12R33 + (15)343(14) + (14)I(14)3 = 103L43
```

with 21 choices for R, I, L with R < I. This, alone, gives 42 solutions while similar results for other values gave the total of 340 solutions!

This next one turned out to be solvable both as an addition and as a subtraction alphametic.

```
SQUARE + DANCE = DANCER   DANCER = 915736
```

```
SQUARE - DANCE = DANCER    DANCER = 574280

GREEN GREY FAWN YELLOW = 89007 8901 4357 102265
```

 The next one was solvable only as a subtraction alphametic (many readers assumed it was an addition alphametic and remarked that it was not solvable.)

```
CANADA - UNITED = STATES   920272 - 408137 = 512135

HAPPY . . . DAYS AHEAD = 29661 . . . 3910 92893
```

Issue 12, December, 1962

```
    MATH      1345
    CLUB      8697
   _____   _____

   MEETS     10042
```

```
ALFRED ÷ E = NEUMAN    704836 ÷ 3 = 231572
```

Harry L. Nelson of Livermore, California, notes that, in base-9

```
    164057 * 5 = 852318
```

```
    ZERO      9635
    ONE         546   with 4 and 8 interchangeable
    TWO         185
   _____   _____

   THREE     10366
```

```
     RMM      599
     HAS      162
   _____   _____

    xxxx     97038
    xxxx
     RMM
   _____

   MOVED
```

```
                    SUN
                   LOSE
                  UNTIE
                 BOTTLE
                ELISION
               NINETEEN
              NONENTITY
             EBULLIENT
            INSOLUBLE
            ─────────
            NEBULOSITY

NEBULOSITY = 1234567890
```

Issue Number 13, February, 1963

```
  *      ANTS        4129
         CANT        5412
         ─────       ─────
         SCAN        9541

         OBEY        3980
         SAM          674
         SAM          674
         ─────       ─────
         VOTE        5328
```

```
 **      XXXX                    XXXX
         XXXX                  XXXXXXX
        ──────                ─────────
        XXXXX                    XXXX
         XXXX                    XXXX
        XXXX                     XXXX
       XXXX                      XXXX
       ──────                   XXXXX
       XXXXXXX                 XXXXX
                              ──────────
                              XXXXXXXXXX
```

$(2348)^2 = 5513104$

$(2348)^3 = 12944768192$

```
***              THIS        4765
                 SURE        5809
                   IS          65
               --------    -------
               PRIME       10639


            FLAT            1072
              AS              73
          --------        -------
            XXXX           78256
            xOLD
          --------
            ACTOR
```

* Some readers did not catch the clue "our ANTS are very odd." If our ANTS could be even, then the only other solution would be ANTS = 1376

** Many readers have asked that the solutions of at least some of the alphametics be shown in some detail. Most alphametics yield a few clues to their solution upon examination of the problem itself. Some simple logical steps – using no pencil and paper – very often narrows the search considerably. The following mental processes are from the Editor's (JSM's) own head. Readers may have approached this particular cryptarithm in a different fashion. Please refer to the set-up shown in the February 1963 **RMM**.

(a) The last digit of the 4-digit number has the greatest value; (b) the first digit must be 1, 2, 3; (c) the second digit must be less than 7; (d) the first and second digits of the square must be greater than 2 – which shows (e) that the first digit of the 4-digit number cannot be a 1 and so must be 2 or 3 – which now shows (f) that the first digit of the square must be greater than 3.

 The above shows how the progressive logical steps tend to "feedback" and gradually refine their own conclusions. The complete solution is found by the application of some physical calculations accompanied by further logical deductions.

*** The clue given meant that PRIME was a prime number. Otherwise, other solutions are possible.

Issue Number 14, January-February, 1964

These solutions were never published in **RMM** because number 14 was the last issue. Therefore, these solutions were generated by the editor of this book.

```
OH )  BLOW  ( GO
       BOB
      _____
       xxx
       xxx
      _____
        NO


49 )  3640  ( 74
      343
     _____
       210
       196
     _____
        14


      CAR
    _____
54 ) WHERE
     xxx
    _____
      xxx
      xRU
     _____
       xxx
       xxx
      _____
       ---
```

```
                257
             _____
    254 )  13878
            108
         _____
            307
            270
          _____
            378
            378
          _____
```

CAUSA = SINE + QUA + NON   10790 = 9264 + 870 + 656

10790 = 9864 + 270 + 656

BASE + BALL = GAMES   7483 + 7455 = 14938

```
        NO
        IF
      _____
       xON
       xIF
      _____
      EVEN
        82
        74
      _____
       328
       574
      _____
      6068
```

TWO + SEVEN + ELEVEN = TWENTY

930 + 58682 + 878682 = 938294

930 + 78682 + 858682 = 938294

EAGER + SEA + EAGLE + RIDES = GALES

13916 + 813 + 13971 + 65018 = 93718

SEE + RMM + SEE = MORE   588 + 922 + 588 = 2098

788 + 922 + 788 = 2498